Periodic Table: A Formula Handbook

N.B. Singh

DEDICATION

To Nature,

I dedicate this book to you, the source of all life. You are my inspiration, my teacher, and my friend.

Thank you for teaching me about the beauty of the world around me. Thank you for showing me the power of the natural world. Thank you for giving me a sense of peace and tranquillity.

I promise to do my part to protect you and your many wonders. I will teach my children about the importance of conservation and sustainability. I will work to make the world a better place for all living things.

Thank you for everything, Nature.

With love,

N.B Singh

Contents

Preface

Welcome to "Periodic Table: A Formula Handbook." This handbook is designed to be a comprehensive guide for students, educators, and enthusiasts interested in delving into the intricate world of chemistry through the lens of the periodic table.

Purpose of the Handbook

The primary purpose of this handbook is to provide a concise and accessible resource for understanding the fundamental concepts, formulas, and applications associated with the periodic table. Whether you are a student navigating the complexities of chemical reactions, an educator seeking supplemental material for your lessons, or an enthusiast eager to explore the wonders of the elements, this handbook aims to be your go-to reference.

Key Features

- **Clear Organization:** The content is organized systematically, following the structure of the periodic table. Each chapter focuses on specific elements, their properties, and their relevance in various chemical processes.

- **Comprehensive Content:** From basic atomic structure to advanced topics like quantum mechanics and nuclear chemistry, this handbook covers a wide range of subjects related to the periodic table.

- **Chemfig Illustrations:** Visual representations, drawn using the `chemfig` package in LaTeX, enhance the understanding of chemical structures, reactions, and bonding.

- **Real-world Applications:** Practical examples, working problems, and numerical exercises are included to demonstrate the real-world applications of chemical principles.

How to Use This Handbook

This handbook is designed to be flexible, allowing readers to navigate chapters based on their interests and requirements. Whether you seek a quick reference or an in-depth exploration of a specific topic, you can easily find the information you need.

Thank you for choosing "Periodic Table: A Formula Handbook." I hope it serves as a valuable companion in your journey through the fascinating world of chemistry.

Chapter 1

Introduction

1.1 Understanding the Periodic Table

The periodic table is a cornerstone of chemistry, offering a systematic arrangement of elements based on their atomic structure and properties. Dmitri Mendeleev's early attempts in 1869 laid the groundwork for this organization, which has evolved into the modern periodic table.

1.1.1 Historical Overview

Mendeleev's periodic table, initially organized by increasing atomic mass, left gaps for undiscovered elements. With the subsequent discovery of the proton and refinement of the atomic theory, the modern periodic table emerged. Now, elements are arranged by atomic number, reflecting the number of protons in an atom.

1.1.2 Modern Periodic Table

The modern periodic table is a visual representation of the periodic law, illustrating the recurring patterns in elemental properties. Periods (rows) and groups (columns) categorize elements. Elements within the same group share similar chemical behavior due to identical valence electron configurations, simplifying predictions of their properties.

1.1.3 Elements and Atomic Structure

The foundation of understanding an element's place on the periodic table lies in its atomic structure. The atomic number dictates the number of protons, uniquely defining an element. Mass number considers both protons and neutrons, distinguishing isotopes – variants of an element with the same atomic number but different mass numbers.

Sample Working Example: Electron Configuration

To determine the electron configuration of oxygen (O), with an atomic number of 8, we follow the Aufbau principle. The electron configuration is $1s^2\ 2s^2\ 2p^4$.

Numerical Example: Isotope Calculation

For hydrogen (H), the two main isotopes are ^1H (protium) and ^2H (deuterium). Given their abundances as 99.985% and 0.015%, respectively, calculate the average atomic mass.

$$\text{Average Atomic Mass} = (1 \times 0.99985) + (2 \times 0.00015)$$
$$= 1.00015\,\text{u} \tag{1.1}$$

Additional Explanation: Trends in the Periodic Table

The periodic table exhibits trends that aid in understanding element behavior. Atomic radius decreases across a period due to increased nuclear charge, while it increases down a group. Electronegativity follows a similar pattern, affecting chemical reactivity.

Practical Insight: Periodic Table and Bonding

The periodic table directly influences chemical bonding. Elements in the same group often form similar compounds, as their outermost electron configurations are comparable. This insight simplifies predicting the types of bonds elements are likely to form.

Chapter 2

Periodic Trends

2.1 Atomic Radius

The atomic radius is a crucial property in the periodic table, representing the size of an atom. It is defined as half the distance between the nuclei of two identical atoms when they are bonded together. The trend of atomic radius across periods and down groups provides valuable insights into the variations in atomic size.

2.1.1 Atomic Radius Across Periods

As one moves across a period from left to right, the atomic radius generally decreases. This trend is attributed to the increasing effective nuclear charge, pulling the electrons closer to the nucleus. The higher positive charge in the nucleus exerts a stronger attractive force on the electrons, resulting in a smaller atomic radius.

$$\text{Atomic Radius} \propto \frac{1}{\text{Effective Nuclear Charge}} \tag{2.1}$$

Sample Working Example: Atomic Radius in Period 3

Consider the atomic radii of elements in Period 3: sodium (Na), magnesium (Mg), aluminum (Al), silicon (Si), phosphorus (P), sulfur (S), chlorine (Cl), and argon (Ar). The trend shows a gradual decrease from left to right.

2.1.2   Atomic Radius Down Groups

In contrast to periods, as one moves down a group, the atomic radius increases. This trend is influenced by the addition of electron shells. Each new shell adds to the overall size of the atom, leading to a larger atomic radius.

$$\text{Atomic Radius} \propto \text{Number of Electron Shells} \tag{2.2}$$

Numerical Example: Atomic Radius in Group 2

Examine the atomic radii of alkaline earth metals in Group 2: beryllium (Be), magnesium (Mg), calcium (Ca), strontium (Sr), and barium (Ba). The trend illustrates an increase in atomic radius down the group.

Additional Insight: Anomalous Trends

While the general trends hold, certain exceptions exist. For instance, transition metals exhibit less regular trends in atomic radius due to variations in electron configurations and the shielding effect.

Practical Application: Atomic Radius in Chemistry

Understanding atomic radius is crucial in predicting chemical behavior. Elements with larger atomic radii are more likely to lose electrons and form cations, while those with smaller atomic radii tend to gain electrons and form anions.

2.2   Ionization Energy

Ionization energy is a crucial property that describes the energy required to remove an electron from an atom or a positive ion. Understanding the trends in ionization energy across the periodic table provides insights into the reactivity and stability of elements.

2.2.1   Ionization Energy Across Periods

Ionization energy generally increases as one moves across a period from left to right. This trend is a result of the increasing effective nuclear charge, which makes it more difficult to remove electrons. The outermost electrons experience a stronger pull from the nucleus, requiring more energy for removal.

$$\text{Ionization Energy} \propto \text{Effective Nuclear Charge} \qquad (2.3)$$

Sample Working Example: Ionization Energy in Period 2

Consider the ionization energies of elements in Period 2: lithium (Li), beryllium (Be), boron (B), carbon (C), nitrogen (N), oxygen (O), fluorine (F), and neon (Ne). The trend shows a gradual increase from left to right.

2.2.2   Ionization Energy Down Groups

In contrast to periods, ionization energy decreases as one moves down a group. This is attributed to the increasing number of electron shells. Electrons in outer shells are farther from the nucleus, experiencing weaker attractive forces, making them easier to remove.

$$\text{Ionization Energy} \propto \frac{1}{\text{Number of Electron Shells}} \qquad (2.4)$$

Numerical Example: Ionization Energy in Group 1

Examine the ionization energies of alkali metals in Group 1: lithium (Li), sodium (Na), potassium (K), rubidium (Rb), and cesium (Cs). The trend illustrates a decrease in ionization energy down the group.

Additional Insight: Anomalous Trends

Similar to atomic radius, certain exceptions exist in the ionization energy trends. For example, elements in transition metals may exhibit variations due to electron configurations.

Practical Application: Ionization Energy in Chemistry

Ionization energy plays a crucial role in chemical reactions, influencing the likelihood of an element forming positive ions (cations). Elements with lower ionization energies are more likely to lose electrons and become cations.

2.3   Electron Affinity

Electron affinity is a key property of elements that describes the energy change when an atom gains an electron to form a negative ion. Understanding the trends in electron affinity across the periodic table provides insights into an element's willingness to accept an additional electron.

2.3.1   Electron Affinity Across Periods

Electron affinity tends to increase across a period from left to right. This trend is influenced by factors such as effective nuclear charge and atomic size. Elements on the right side of the periodic table have higher electron affinities as they strive to achieve a stable electron configuration.

$$\text{Electron Affinity} \propto \text{Effective Nuclear Charge} \tag{2.5}$$

Sample Working Example: Electron Affinity in Period 3

Consider the electron affinities of elements in Period 3: sodium (Na), magnesium (Mg), aluminum (Al), silicon (Si), phosphorus (P), sulfur (S), chlorine (Cl), and argon (Ar). The trend shows an increase from left to right.

2.3.2   Electron Affinity Down Groups

In contrast to periods, electron affinity generally decreases as one moves down a group. Elements in lower periods have larger atomic sizes, making it less favorable for them to gain electrons. Additionally, the increased electron shielding reduces the effective nuclear charge.

$$\text{Electron Affinity} \propto \frac{1}{\text{Atomic Size}} \tag{2.6}$$

Numerical Example: Electron Affinity in Group 17

Examine the electron affinities of halogens in Group 17: fluorine (F), chlorine (Cl), bromine (Br), iodine (I), and astatine (At). The trend illustrates a decrease in electron affinity down the group.

Additional Insight: Anomalous Trends

While the general trends hold, there are exceptions due to factors like electron repulsion and atomic stability. For example, noble gases have very low electron affinities.

Practical Application: Electron Affinity in Chemistry

Knowledge of electron affinity is crucial in understanding chemical reactions, especially those involving the formation of negative ions (anions). Elements with higher electron affinities are more likely to accept electrons and form stable anions.

2.4 Electronegativity

Electronegativity is a fundamental property that characterizes an element's ability to attract electrons in a chemical bond. It plays a pivotal role in understanding the nature of chemical compounds and predicting the type of bonding that occurs.

2.4.1 Electronegativity Across Periods

The electronegativity of elements generally increases across a period from left to right in the periodic table. This trend is influenced by factors such as effective nuclear charge and atomic size. Elements on the right side of the table have higher electronegativities as they tend to attract electrons more strongly.

$$\text{Electronegativity} \propto \text{Effective Nuclear Charge} \tag{2.7}$$

Sample Working Example: Electronegativity in Period 2

Consider the electronegativities of elements in Period 2: lithium (Li), beryllium (Be), boron (B), carbon (C), nitrogen (N), oxygen (O), fluorine (F), and neon (Ne). The trend shows a gradual increase from left to right.

2.4.2   Electronegativity Down Groups

Contrary to periods, electronegativity generally decreases as one moves down a group. Elements in lower periods have larger atomic sizes, making it less favorable for them to attract electrons. Additionally, the increased electron shielding reduces the effective nuclear charge.

$$\text{Electronegativity} \propto \frac{1}{\text{Atomic Size}} \tag{2.8}$$

Numerical Example: Electronegativity in Group 17

Examine the electronegativities of halogens in Group 17: fluorine (F), chlorine (Cl), bromine (Br), iodine (I), and astatine (At). The trend illustrates a decrease in electronegativity down the group.

Additional Insight: Anomalous Trends

While the general trends hold, there are exceptions due to factors like electron repulsion and atomic stability. Noble gases, for instance, have very low electronegativities.

Practical Application: Electronegativity in Chemistry

Electronegativity is a key determinant in the type of chemical bonding that occurs between elements. Elements with higher electronegativities are more likely to attract electrons and form polar or ionic bonds in compounds.

Chapter 3

Chemical Families

3.1 Alkali Metals

The alkali metals, found in Group 1 of the periodic table, are a group of highly reactive elements. This section explores their distinctive characteristics, properties, and chemical behaviors.

3.1.1 General Characteristics

Alkali metals include lithium (Li), sodium (Na), potassium (K), rubidium (Rb), cesium (Cs), and francium (Fr). These metals share common features such as low density, low melting points, and a single valence electron. They are highly reactive and tend to form positive ions (+) by losing this valence electron.

Electronic Configuration

The electronic configuration of alkali metals involves a single valence electron in their outermost electron shell.

$$\text{Li: } 1s^2 2s^1, \quad \text{Na: } 1s^2 2s^2 2p^6 3s^1, \quad \text{K: } 1s^2 2s^2 2p^6 3s^2 3p^6 4s^1$$

3.1.2   Chemical Properties

Alkali metals are known for their vigorous reactions with water and air. The reactivity increases down the group, with cesium being the most reactive alkali metal.

Numerical Example: Reactivity Trends

Compare the reactivity of lithium, sodium, and potassium. The reactivity increases down the group, with potassium being more reactive than sodium, and sodium more reactive than lithium.

3.1.3   Applications

Alkali metals find applications in various industries. Sodium, for example, is used in the production of soap, while potassium is essential for fertilizers. Lithium is widely used in batteries due to its lightweight and high electrochemical potential.

3.2   Transition Metals

The transition metals, situated in the central blocks of the periodic table, are a group of elements with unique properties. This section explores their characteristics, electronic configurations, and applications.

3.2.1   General Characteristics

Transition metals include elements like iron (Fe), copper (Cu), and gold (Au). They are known for their variable oxidation states, high melting points, and the ability to form colorful compounds.

Electronic Configuration

The electronic configuration of transition metals is characterized by the filling of the d orbitals. For example:

$$\text{Fe: } 1s^2 2s^2 2p^6 3s^2 3p^6 4s^2 3d^6, \quad \text{Cu: } 1s^2 2s^2 2p^6 3s^2 3p^6 4s^2 3d^9$$

3.2.2 Physical and Chemical Properties

Transition metals exhibit high melting and boiling points, good conductivity, and magnetic properties. They are known for forming complex ions and compounds with a wide range of coordination numbers.

Example: Iron (Fe)

Iron is a crucial transition metal with various applications. Its ability to form different oxidation states makes it essential in biological systems, as in the heme group of hemoglobin.

3.2.3 Applications

Transition metals find applications in catalysis, electronics, and medicine. For instance, platinum (Pt) is widely used as a catalyst in chemical reactions, and copper (Cu) is employed in electrical wiring.

Numerical Example: Catalytic Activity

Compare the catalytic activity of platinum and palladium. Platinum, with its ability to adsorb hydrogen effectively, shows higher catalytic activity in certain reactions compared to palladium.

3.3 Halogens

The halogens, situated in Group 17 of the periodic table, are a group of highly reactive nonmetals. This section explores their characteristics, electronic configurations, and chemical behaviors.

3.3.1   General Characteristics

Halogens include fluorine (F), chlorine (Cl), bromine (Br), iodine (I), and astatine (At). They are known for their high electronegativity and tendency to form salts by gaining one electron.

Electronic Configuration

The electronic configuration of halogens involves the filling of the p orbitals. For example:

$$\text{F: } 1s^2 2s^2 2p^5, \quad \text{Cl: } 1s^2 2s^2 2p^6 3s^2 3p^5$$

3.3.2   Physical and Chemical Properties

Halogens exhibit distinct physical states across the group, from gases (fluorine, chlorine) to a liquid (bromine) and a solid (iodine). They are highly reactive and readily form diatomic molecules.

Example: Chlorine (Cl_2)

Chlorine is a widely used halogen, known for its disinfectant properties. It readily forms Cl_2 molecules, which are effective in water treatment and sanitation.

3.3.3   Chemical Reactions

Halogens participate in various chemical reactions, forming compounds with metals and nonmetals. For example:

- $2\,Na\,(s) + Cl_2(g) \longrightarrow 2\,NaCl\,(s)$
- $H_2(g) + F_2(g) \longrightarrow 2\,HF\,(g)$

Numerical Example: Reactivity Series

Compare the reactivity of fluorine and iodine. Fluorine, being the most electronegative element, is more reactive than iodine.

3.3.4 Applications

Halogens find applications in various industries. Chlorine is used in water treatment, fluorine in the production of fluorinated compounds, and iodine in medical applications.

3.4 Noble Gases

The noble gases, found in Group 18 of the periodic table, are a group of inert and colorless gases. This section explores their unique characteristics, electronic configurations, and applications.

3.4.1 General Characteristics

Noble gases include helium (He), neon (Ne), argon (Ar), krypton (Kr), xenon (Xe), and radon (Rn). They are characterized by their complete outer electron shells, making them chemically inert.

Electronic Configuration

The electronic configuration of noble gases is characterized by a fully-filled outer shell. For example:

$$\text{He: } 1s^2, \quad \text{Ne: } 1s^2 2s^2 2p^6, \quad \text{Xe: } 1s^2 2s^2 2p^6 3s^2 3p^6 4s^2 3d^{10} 4p^6 5s^2 4d^{10} 5p^6$$

3.4.2 Physical and Chemical Properties

Noble gases are colorless, odorless, and tasteless. They have low boiling points and are monatomic gases under normal conditions. Due to their inert nature, they rarely form compounds.

Example: Helium (He)

Helium is widely used in various applications, including as a lifting gas in balloons and as a coolant in certain scientific and medical equipment.

3.4.3 Applications

Noble gases find applications in different fields. Helium is used in cryogenics, neon in neon signs, and argon in welding.

Numerical Example: Neon Signs

Examine the use of neon in producing colorful signs. When an electric current passes through neon gas, it emits a characteristic glow, resulting in the vibrant colors seen in neon signs.

3.5   Lanthanides and Actinides

The lanthanides and actinides, often referred to as the rare earth elements and the actinide series, are a group of elements located in the f-block of the periodic table. This section explores their unique electronic configurations, properties, and applications.

3.5.1 Lanthanides

The lanthanides consist of 15 elements from atomic number 57 (lanthanum, La) to 71 (lutetium, Lu). They are known for their similar properties and the filling of the 4f orbital.

Electronic Configuration

The electronic configuration of lanthanides involves the filling of the 4f orbital. For example:

$$Eu: 1s^2 2s^2 2p^6 3s^2 3p^6 4s^2 3d^{10} 4p^6 5s^2 4d^{10} 5p^6 6s^2 4f^7$$

3.5.2   Actinides

The actinides consist of 15 elements from atomic number 89 (actinium, Ac) to 103 (lawrencium, Lr). They are characterized by the filling of the 5f orbital.

Electronic Configuration

The electronic configuration of actinides involves the filling of the 5f orbital. For example:

$$\text{Th:}\ 1s^2 2s^2 2p^6 3s^2 3p^6 4s^2 3d^{10} 4p^6 5s^2 4d^{10} 5p^6 6s^2 4f^{14} 5d^{10} 6p^6 7s^2 5f^0$$

3.5.3   Properties

Both lanthanides and actinides share common properties, including the ability to form colorful compounds and exhibit variable oxidation states.

Example: Europium (Eu)

Europium is a lanthanide known for its ability to exhibit different oxidation states and emit red fluorescence when exposed to ultraviolet light.

3.5.4   Applications

Lanthanides and actinides find applications in various technologies. Gadolinium (Gd) is used in magnetic resonance imaging (MRI), while uranium (U) is essential in nuclear power generation.

Numerical Example: Uranium Enrichment

Explore the process of uranium enrichment for nuclear reactors. By increasing the concentration of uranium-235, reactors can efficiently generate nuclear energy.

Chapter 4

Elemental Properties and Formulas

4.1 Hydrogen and Helium

The first two elements in the periodic table, hydrogen (H) and helium (He), are unique in their properties and play essential roles in various processes. This section explores their characteristics, electronic configurations, and applications.

4.1.1 Hydrogen (H)

Hydrogen is the lightest and most abundant element in the universe. It has the simplest atomic structure with only one electron.

Electronic Configuration

The electronic configuration of hydrogen is given by:

$$\text{H: } 1s^1$$

4.1.2   Helium (**He**)

Helium is a noble gas known for its inert nature and application in various fields, particularly as a cooling agent.

Electronic Configuration

The electronic configuration of helium is given by:

$$\text{He: } 1s^2$$

4.1.3   Properties

Hydrogen exhibits diatomic behavior (H_2), forming molecular hydrogen under normal conditions. Helium, being a noble gas, is colorless, odorless, and inert.

Example: Hydrogen Gas

Explore the properties of hydrogen gas (H_2). It is flammable and is used as a fuel in various industrial applications.

4.1.4   Applications

Both hydrogen and helium find applications in diverse fields. Hydrogen is used in fuel cells, while helium is crucial in cryogenics and as a lifting gas.

Numerical Example: Hydrogen Fuel Cells

Discuss the use of hydrogen fuel cells in powering vehicles. Analyze the efficiency and environmental benefits of hydrogen as a clean fuel source.

4.2   Groups 1 and 2 Elements

The elements in Groups 1 and 2 of the periodic table, known as alkali metals and alkaline earth metals, respectively, exhibit unique properties and play essential roles in various chemical reactions. This section explores their characteristics, electronic configurations, and applications.

4.2.1 Alkali Metals (Group 1)

Alkali metals include lithium (Li), sodium (Na), potassium (K), rubidium (Rb), cesium (Cs), and francium (Fr). They are highly reactive and tend to form +1 oxidation states.

Electronic Configuration

The electronic configuration of alkali metals involves the filling of the outermost s orbital. For example:

$$\text{K: } 1s^2 2s^2 2p^6 3s^2 3p^6 4s^1$$

4.2.2 Alkaline Earth Metals (Group 2)

Alkaline earth metals include beryllium (Be), magnesium (Mg), calcium (Ca), strontium (Sr), barium (Ba), and radium (Ra). They are less reactive than alkali metals and tend to form +2 oxidation states.

Electronic Configuration

The electronic configuration of alkaline earth metals involves the filling of the outermost s and p orbitals. For example:

$$\text{Ca: } 1s^2 2s^2 2p^6 3s^2 3p^6 4s^2$$

4.2.3 Properties

Alkali metals are soft, highly reactive metals that react vigorously with water. Alkaline earth metals are harder and less reactive, but still form basic oxides and hydroxides.

Example: Reaction of Sodium with Water

Explore the reaction of sodium with water:

$$2\,Na\,(s) + 2\,H_2O\,(l) \longrightarrow 2\,NaOH\,(aq) + H_2(g)$$

4.2.4   Applications

Alkali and alkaline earth metals find applications in various industries. Sodium is used in food preparation, and calcium is essential for bone health.

Numerical Example: Calcium Supplementation

Discuss the importance of calcium supplementation for individuals with calcium deficiency. Analyze the chemical processes involved in calcium absorption.

4.3   Transition Metal Elements

The transition metals, occupying the d-block of the periodic table, exhibit unique properties and versatile oxidation states. This section explores the characteristics, electronic configurations, and applications of transition metals.

4.3.1   Electronic Configurations

The electronic configurations of transition metals involve the filling of the d-block orbitals. For example:

$$\text{Fe: } 1s^2 2s^2 2p^6 3s^2 3p^6 4s^2 3d^6 4p^6 5s^2 4d^6$$

4.3.2   Versatile Oxidation States

Transition metals are known for their ability to exhibit multiple oxidation states. For example, iron (Fe) can exist in $+2$ and $+3$ oxidation states.

Example: Iron Oxides

Explore the formation of iron oxides:

$$4\,Fe(s) + 3\,O_2(g) \longrightarrow 2\,Fe_2O_3(s)$$

4.3.3 Properties

Transition metals have high melting and boiling points, good conductivity, and form colorful compounds. They are often used as catalysts in various chemical reactions.

Numerical Example: Catalytic Activity of Platinum

Discuss the catalytic activity of platinum (Pt) in the hydrogenation of unsaturated hydrocarbons. Analyze the chemical processes involved.

4.3.4 Applications

Transition metals find applications in numerous industries, including medicine, electronics, and catalysis.

Example: Copper (Cu) in Electronics

Explore the use of copper in electronic devices, emphasizing its conductivity and applications in wiring.

4.4 Nonmetals and Metalloids

Nonmetals and metalloids form a diverse group of elements with distinct properties. This section delves into the characteristics, electronic configurations, and applications of nonmetals and metalloids.

4.4.1   Nonmetals

Nonmetals, found in the p-block of the periodic table, have varied properties. Elements like oxygen (O) and nitrogen (N) play crucial roles in biological processes.

Electronic Configuration

The electronic configuration of nonmetals involves the filling of the p-block orbitals. For example:

$$\text{O: } 1s^2 2s^2 2p^4$$

4.4.2   Metalloids

Metalloids, situated between metals and nonmetals, exhibit properties of both. Boron (B) and silicon (Si) are examples of metalloids.

Example: Silicon

Explore the electronic configuration and properties of silicon:

$$\text{Si: } 1s^2 2s^2 2p^6 3s^2 3p^6 4s^2 3d^{10} 4p^2$$

4.4.3   Properties

Nonmetals are generally poor conductors of heat and electricity, while metalloids show intermediate conductivity. They are essential in various chemical reactions and biological processes.

Example: Combustion of Hydrogen

Examine the combustion reaction of hydrogen (H_2):

$$2\,H_2(g) + O_2(g) \longrightarrow 2\,H_2O\,(l)$$

4.4.4 Applications

Nonmetals and metalloids find applications in diverse fields. Silicon is crucial in the semiconductor industry, and oxygen is essential for respiration.

Numerical Example: Semiconductor Devices

Discuss the use of silicon in semiconductor devices. Explore the role of doping in altering its conductivity.

4.5 Rare Earth Elements

The rare earth elements, located in the lanthanide and actinide series, possess unique properties and applications. This section explores the characteristics, electronic configurations, and uses of rare earth elements.

4.5.1 Lanthanide Series

The lanthanide series consists of 15 elements, from cerium (Ce) to lutetium (Lu). These elements share similar properties, with electron configurations that fill the 4f orbitals.

Electronic Configuration

The electronic configuration of lanthanides involves the filling of the 4f orbitals. For example:

$$\text{Eu: } 1s^2 2s^2 2p^6 3s^2 3p^6 4s^2 3d^{10} 4p^6 5s^2 4d^{10} 5p^6 6s^2 4f^7$$

4.5.2 Actinide Series

The actinide series, found below the lanthanides, includes elements such as uranium (U) and thorium (Th). These elements exhibit similar properties and have 5f orbitals in their electronic configurations.

Example: Uranium

Explore the electronic configuration and properties of uranium:

$$U: 1s^2 2s^2 2p^6 3s^2 3p^6 4s^2 3d^{10} 4p^6 5s^2 4d^{10} 5p^6 6s^2 4f^{14} 5d^{10} 6p^6 7s^2 5f^3$$

4.5.3   Properties

Rare earth elements exhibit unique magnetic, optical, and catalytic properties. They are vital components in various technological applications.

Example: Neodymium Magnets

Discuss the use of neodymium (Nd) magnets in modern technology, emphasizing their strong magnetic properties.

4.5.4   Applications

Rare earth elements find applications in the production of electronics, catalysts, and medical imaging.

Numerical Example: Gadolinium in MRI Contrast Agents

Explore the use of gadolinium (Gd) in magnetic resonance imaging (MRI) contrast agents. Discuss the enhancement of imaging quality.

Chapter 5

Chemical Bonding

5.1   Ionic Bonds

Ionic bonds are formed through the transfer of electrons between atoms, leading to the creation of ions. This section delves into the principles of ionic bonding, its characteristics, and various examples.

5.1.1   Principles of Ionic Bonding

In ionic bonding, electrons are transferred from one atom to another, resulting in the formation of positively charged cations and negatively charged anions.

Example: Sodium Chloride ($NaCl$)

The formation of sodium chloride involves the transfer of an electron from sodium (Na) to chlorine (Cl), leading to the creation of Na^+ and Cl^- ions:

$$Na + Cl \longrightarrow Na^+ + Cl^-$$

5.1.2   Characteristics of Ionic Compounds

Ionic compounds exhibit specific properties, including high melting points, electrical conductivity in molten or dissolved states, and solubility in water.

Example: Conductivity of Molten Sodium Chloride

Molten sodium chloride conducts electricity due to the movement of ions, demonstrating one of the characteristic features of ionic compounds.

5.1.3   Applications of Ionic Compounds

Ionic compounds find applications in various industries, such as the production of salts, pharmaceuticals, and ceramics.

Numerical Example: Electrolysis of Sodium Chloride

Explore the electrolysis of sodium chloride to produce sodium hydroxide and chlorine gas.

5.1.4   Ionic Bonding in Biological Systems

Ionic bonding plays a crucial role in biological processes, influencing the structure and function of biomolecules.

Example: Calcium Ions in Bone Formation

Discuss the role of calcium ions in the ionic bonding within hydroxyapatite, a key component of bone structure.

5.2   Covalent Bonds

Covalent bonds involve the sharing of electrons between atoms, resulting in the formation of molecules. This section explores the principles of covalent bonding, its characteristics, and provides various examples.

5.2.1 Principles of Covalent Bonding

In covalent bonding, electrons are shared between atoms to achieve a stable electron configuration. This sharing creates a bond between the atoms.

Example: Hydrogen Molecule (H_2)

The covalent bond in a hydrogen molecule involves the sharing of electrons between two hydrogen atoms:

$$H + H \longrightarrow H_2$$

5.2.2 Characteristics of Covalent Compounds

Covalent compounds exhibit specific properties, including lower melting points, non-conductivity in most cases, and solubility in non-polar solvents.

Example: Properties of Oxygen (O_2)

Oxygen, in its diatomic form (O_2), is a covalent molecule with distinct characteristics, such as its role in supporting combustion.

5.2.3 Types of Covalent Bonds

Covalent bonds can be classified into polar and nonpolar bonds based on the electronegativity difference between atoms.

Numerical Example: Electronegativity in HCl

Calculate the electronegativity difference in hydrogen chloride (HCl) and determine the type of covalent bond formed.

5.2.4 Multiple Covalent Bonds

Some molecules involve the sharing of multiple pairs of electrons, leading to double or triple covalent bonds.

Example: Carbon Dioxide (CO_2)

Explore the structure of carbon dioxide, highlighting the double covalent bonds between carbon and oxygen atoms.

5.2.5   Applications of Covalent Compounds

Covalent compounds play crucial roles in various applications, including the pharmaceutical industry, polymers, and organic chemistry.

Example: Polymerization of Ethene

Discuss the covalent bonding in the polymerization process of ethene to form polyethylene.

5.2.6   Covalent Bonding in Biological Molecules

The structure of biological molecules often involves covalent bonds, influencing their function and properties.

Example: DNA Double Helix

Examine the covalent bonds within the DNA double helix, emphasizing the importance of hydrogen bonding.

5.3   Metallic Bonds

Metallic bonding is a unique type of bonding that occurs in metals, leading to their characteristic properties. This section explores the principles of metallic bonding and its applications.

5.3.1   Principles of Metallic Bonding

In metallic bonding, electrons are delocalized and free to move within the structure of a metal, creating a "sea of electrons" that contributes to the metal's properties.

Example: Sodium (Na) Metal

The metallic bonding in sodium involves the delocalization of electrons, creating a conductive and malleable material.

5.3.2 Characteristics of Metallic Compounds

Metallic compounds exhibit several distinctive characteristics, including conductivity, malleability, ductility, and luster.

Numerical Example: Electrical Conductivity of Copper (Cu)

Explore the relationship between metallic bonding and the exceptional electrical conductivity of copper.

5.3.3 Alloys and Solid Solutions

Alloys are materials composed of two or more elements, often metals, with metallic bonding contributing to their unique properties.

Example: Steel

Discuss the metallic bonding in steel, a widely used alloy with enhanced strength and durability.

5.3.4 Band Theory and Metallic Bonding

Band theory explains the electronic structure of metals and the formation of energy bands.

Example: Band Structure of Silver (Ag)

Illustrate the band structure of silver and its correlation with the metal's properties.

5.3.5   Applications of Metallic Bonds

Metallic bonds play a crucial role in various applications, from electrical conductors to structural materials.

Example: Aluminum in Aircraft Construction

Examine how metallic bonding in aluminum contributes to its lightweight and strong properties, making it suitable for aircraft construction.

5.3.6   Metallic Bonding in Biology

While metallic bonding is not common in biological molecules, certain biological processes involve metal ions with unique bonding properties.

Example: Hemoglobin

Explore the role of metallic bonding in hemoglobin, where iron ions play a crucial role in oxygen transport.

5.4   Bond Polarity

Bond polarity is a crucial aspect of chemical bonding, determining the distribution of electrons between atoms and influencing the overall behavior of molecules.

5.4.1   Electronegativity and Bond Polarity

Electronegativity, a measure of an atom's ability to attract electrons, plays a key role in determining bond polarity.

Definition of Electronegativity

Electronegativity is defined as the tendency of an atom to attract a bonding pair of electrons.

Numerical Example: Electronegativity Difference in HCl

Explore the electronegativity difference between hydrogen (H) and chlorine (Cl) in hydrochloric acid (HCl).

5.4.2 Polarity of Covalent Bonds

Covalent bonds can be nonpolar or polar, depending on the electronegativity difference between the atoms involved.

Example: Polar H_2O Molecule

Examine the polar nature of the water molecule due to the electronegativity difference between oxygen and hydrogen.

5.4.3 Dipole Moments

Dipole moments are vectors that indicate the direction and magnitude of the charge separation in a molecule.

Calculation of Dipole Moment

Learn how to calculate the dipole moment of a molecule based on the bond polarity and molecular geometry.

Example: Dipole Moment in CO_2

Explore the dipole moment in carbon dioxide (CO_2) and its relationship to the molecule's linear structure.

5.4.4 Effect of Molecular Geometry on Polarity

The three-dimensional arrangement of atoms in a molecule influences its overall polarity.

Example: Polar CH3Cl Molecule

Investigate the molecular geometry of chloromethane and its impact on the molecule's polarity.

5.4.5 Polarity in Ionic Bonds

Ionic bonds also exhibit polarity, with the more electronegative element attracting electrons more strongly.

Example: Polarity in NaCl

Discuss the polarity in sodium chloride (NaCl) and how it contributes to the compound's properties.

5.4.6 Applications of Understanding Bond Polarity

Understanding bond polarity is essential in various fields, including chemistry, biology, and materials science.

Example: Drug Design and Molecular Polarity

Explore how knowledge of bond polarity is applied in drug design, considering the interaction between drugs and biological molecules.

5.5 Lewis Structures

Lewis structures provide a visual representation of how valence electrons are distributed in a molecule or ion, offering insights into its bonding and geometry.

5.5.1 Basic Principles of Lewis Structures

Lewis structures follow specific rules for the placement of electrons around atoms:

- Atoms are represented by their chemical symbols.

- Dots or lines are used to represent valence electrons.

- Each bond (single, double, or triple) is represented by a pair of electrons or lines.

- Octet rule: Atoms strive to achieve a full outer shell (eight electrons).

5.5.2 Drawing Lewis Structures

The process of drawing Lewis structures involves several steps:

Counting Valence Electrons

Determine the total number of valence electrons for all atoms in the molecule.

Placing Electrons Around Atoms

Distribute the electrons around the atoms, following the octet rule. Hydrogen usually forms only one bond.

Forming Bonds

Connect atoms with single, double, or triple bonds until each atom satisfies the octet rule.

Checking Formal Charges

Verify that formal charges are minimized. A Lewis structure with the lowest formal charges is preferred.

Example: Lewis Structure of H_2O

Illustrate the step-by-step process of drawing the Lewis structure for a water molecule.

5.5.3   Resonance Structures

Some molecules can have multiple Lewis structures, known as resonance structures, that contribute to the actual structure.

Example: Resonance in O_3

Explore the resonance structures of ozone (O_3) and their implications for the molecule's stability.

5.5.4   Exceptions to the Octet Rule

Certain molecules and ions deviate from the octet rule, displaying expanded or incomplete octets.

Example: Expanded Octet in SF_6

Examine the Lewis structure of sulfur hexafluoride (SF_6) and its deviation from the octet rule.

5.5.5   Lewis Structures for Polyatomic Ions

Polyatomic ions are treated similarly to molecules when drawing Lewis structures.

Example: Lewis Structure of NO_3^-

Demonstrate the process of drawing the Lewis structure for the nitrate ion (NO_3^-).

5.5.6   Applications of Lewis Structures

Understanding Lewis structures is crucial in predicting molecular geometry, reactivity, and intermolecular forces.

Example: Predicting Molecular Shape in CH4

Discuss how the Lewis structure of methane (CH_4) helps predict its tetrahedral molecular shape.

Chapter 6

Common Compounds and Formulas

6.1 Oxides and Hydroxides

Oxides and hydroxides are pivotal compounds with diverse applications and essential roles in chemical processes. This section delves into the properties, structures, and applications of selected oxides and hydroxides, offering both theoretical insights and practical considerations.

6.1.1 Water (H_2O)

Water, with the chemical formula H_2O, is a fundamental compound that sustains life and fuels industrial processes. Represented as:

Water's unique properties, such as high surface tension and solvent capabilities, make it indispensable for various applications, from biological systems to industrial cooling processes.

6.1.2 Carbon Dioxide (CO_2)

Carbon dioxide (CO_2) is a vital component in the carbon cycle, produced through various combustion processes. Its structural representation is:

$$O = C = O$$

Apart from its role in photosynthesis, carbon dioxide contributes to the greenhouse effect, impacting Earth's climate.

6.1.3 Calcium Oxide (CaO)

Calcium oxide, or quicklime, forms through the reaction of calcium with oxygen. The chemical equation is:

$$Ca + O_2 \longrightarrow CaO$$

The structural formula of calcium oxide is:

$$
\begin{array}{c}
O^{2-} \\
\diagup \\
Ca^{2+} \\
\diagdown \\
O^{2-}
\end{array}
$$

Utilized in various industries, including cement production, calcium oxide plays a crucial role in shaping modern infrastructure.

6.1.4 Sodium Hydroxide ($NaOH$)

Sodium hydroxide, or caustic soda, results from the reaction of sodium with water:

$$2\,Na + 2\,H_2O \longrightarrow 2\,NaOH + H_2$$

The structural formula of sodium hydroxide is:

$$O^2 \text{———}$$
$$Na^+$$
$$H$$

Widely employed in soap and detergent production, sodium hydroxide is a key component in various chemical processes.

6.2 Acids and Bases

In this section, we explore the properties and formulas of acids and bases. Acids are substances that can donate protons (H^+), while bases are substances that can accept protons.

6.2.1 Acids

The general formula for an acid is represented as HA, where H is the hydrogen ion. An example of a common acid is hydrochloric acid (HCl).

Structure of Acids

The structure of acids can vary, but in the case of hydrochloric acid, it consists of a hydrogen ion bonded to a chloride ion.

$$Cl \text{———}$$
$$H^+$$

6.2.2 Bases

Bases are substances that can accept protons. The general formula for a base is BOH, where B is the base component. An example of a common base is sodium hydroxide (NaOH).

Structure of Bases

The structure of bases involves the base component bonded to a hydroxide ion.

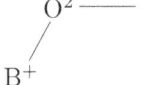

6.2.3   Acid-Base Reactions

Acid-base reactions involve the transfer of protons from an acid to a base. The general form of an acid-base reaction is:

$$HA + BOH \longrightarrow BA + H_2O$$

Example Reaction

Let's consider the reaction between hydrochloric acid and sodium hydroxide:

$$HCl + NaOH \longrightarrow NaCl + H_2O$$

6.2.4   Examples and Applications

- **Example 1: Citric Acid** ($C_6H_8O_7$)

  Citric acid is a weak acid found in citrus fruits. Its chemical formula reflects its molecular composition.

- **Example 2: Ammonium Hydroxide** (NH_4OH)

  Ammonium hydroxide is an example of a base. Its chemical formula indicates the presence of the ammonium ion and the hydroxide ion.

6.3   Salts and Coordination Compounds

In this section, we explore salts and coordination compounds, delving into their structures, formulas, and properties.

6.3.1 Salts

Salts are formed through the reaction of an acid with a base. The general formula for a salt is AB, where A is the cation and B is the anion.

Structure of Salts

The structure of salts depends on the combination of cations and anions. For example, sodium chloride (NaCl) has a simple structure with sodium ions (Na^+) and chloride ions (Cl^-).

$$Cl \text{------}$$
$$/$$
$$Na^+$$

6.3.2 Coordination Compounds

Coordination compounds involve the coordination of a central metal atom or ion with surrounding ligands. The general formula is $[M(L)_n]$, where M is the metal, L is the ligand, and n is the coordination number.

Structure of Coordination Compounds

The structure of coordination compounds varies widely. One example is $[Fe(CN)_6]^{3-}$, where iron (Fe) is coordinated with six cyanide ligands (CN^-).

$$C \equiv N$$
$$\backslash$$
$$Fe^{3+} - C \equiv N$$
$$C \equiv N \equiv N$$

6.3.3 Examples and Applications

- **Example 1: Sodium Chloride (NaCl)**

  Sodium chloride is a common salt used in various applications, including as a seasoning in food.

- **Example 2: Hemoglobin** ($[Fe(H_2O)_6]^{2+}$)

  Hemoglobin, a vital component of blood, contains iron coordinated with water ligands.

Chapter 7

Chemical Reactions

7.1 Types of Chemical Reactions

Chemical reactions can be classified into various types based on the nature of reactants and products. Each type has distinct characteristics and often involves specific reaction mechanisms.

7.1.1 Combination Reactions

Combination reactions involve the union of two or more substances to form a single product. The general form is:

$$A + B \longrightarrow AB$$

For example, the combination of hydrogen and oxygen to form water:

$$2 H_2 + O_2 \longrightarrow 2 H_2O$$

7.1.2 Decomposition Reactions

Decomposition reactions involve the breakdown of a single reactant into two or more products. The general form is:

$$AB \longrightarrow A + B$$

For example, the decomposition of water into hydrogen and oxygen:

$$2H_2O \longrightarrow 2H_2 + O_2$$

7.1.3 Displacement Reactions

Displacement reactions involve the replacement of one element in a compound by another element. The general form is:

$$A + BC \longrightarrow AC + B$$

For example, the displacement of copper from copper sulfate by zinc:

$$Zn + CuSO_4 \longrightarrow ZnSO_4 + Cu$$

7.1.4 Redox Reactions

Redox (oxidation-reduction) reactions involve the transfer of electrons between reactants. Oxidation is the loss of electrons, while reduction is the gain of electrons. The general form is:

$$A + B \longrightarrow A^+ + B^-$$

For example, the reaction between sodium and chlorine:

$$2Na + Cl_2 \longrightarrow 2NaCl$$

7.1.5 Acid-Base Reactions

Acid-base reactions involve the transfer of protons (H+ ions) between reactants. The general form is:

$$HA + BOH \longrightarrow BA + H_2O$$

For example, the reaction between hydrochloric acid (HCl) and sodium hydroxide ($NaOH$):

$$HCl + NaOH \longrightarrow NaCl + H_2O$$

7.1.6 Precipitation Reactions

Precipitation reactions involve the formation of an insoluble solid (precipitate) from the mixing of two solutions. The general form is:

$$AB(aq) + CD(aq) \longrightarrow AD(s) + BC(aq)$$

For example, the reaction between silver nitrate ($AgNO_3$) and sodium chloride ($NaCl$):

$$AgNO_3(aq) + NaCl(aq) \longrightarrow AgCl(s) + NaNO_3(aq)$$

7.1.7 Sample Working Example

Consider the decomposition of hydrogen peroxide (H_2O_2) into water and oxygen:

$$2H_2O_2 \longrightarrow 2H_2O + O_2$$

To find the moles of oxygen produced when 3 moles of hydrogen peroxide decompose, we use stoichiometry:

$$\text{Moles of oxygen} = \text{Moles of hydrogen peroxide} \times \frac{\text{Coefficient of oxygen}}{\text{Coefficient of hydrogen peroxide}}$$

Substituting values:

$$\text{Moles of oxygen} = 3\,\text{mol} \times \frac{1}{2} = 1.5\,\text{mol}$$

7.1.8   Numerical Example

Let's consider the displacement reaction between zinc (Zn) and copper sulfate ($CuSO_4$):

$$Zn + CuSO_4 \longrightarrow ZnSO_4 + Cu$$

If 5 moles of zinc react, we can find the moles of copper formed:

$$\text{Moles of copper} = 5\,\text{mol} \times \frac{1}{1} = 5\,\text{mol}$$

7.2   Balancing Chemical Equations

Balancing chemical equations is a fundamental skill in chemistry. It ensures that the number of atoms of each element on the reactant side is equal to the number of atoms on the product side. Let's explore the process of balancing equations through examples.

7.2.1   Example 1: Combustion of Methane

Consider the combustion of methane (CH_4) in the presence of oxygen (O_2) to produce carbon dioxide (CO_2) and water (H_2O). The unbalanced equation is:

$$CH_4 + O_2 \longrightarrow CO_2 + H_2O$$

To balance the equation, follow these steps:

- Start by balancing carbon (C) and hydrogen (H) atoms.

- Balance oxygen (O) atoms last, using coefficients.

The balanced equation is:

$$CH_4 + 2\,O_2 \longrightarrow CO_2 + 2\,H_2O$$

This ensures that there is one carbon atom on each side and four hydrogen and four oxygen atoms on each side.

7.2.2 Example 2: Reaction Between Iron and Oxygen

Consider the reaction between iron (Fe) and oxygen (O_2) to form iron(III) oxide (Fe_2O_3). The unbalanced equation is:

$$Fe + O_2 \longrightarrow Fe_2O_3$$

Balancing the equation:

$$4\,Fe + 3\,O_2 \longrightarrow 2\,Fe_2O_3$$

This ensures that there are four iron atoms on each side and six oxygen atoms on each side.

7.2.3 Bonding Diagrams

Bonding diagrams visually represent the arrangement of atoms and bonds in a molecule. Let's use to draw the bonding diagram for methane (CH_4):

$$H \!-\!-\! C \!-\!-\! H$$
$$\mid$$
$$H$$

In this diagram, each line represents a bond, and the angles indicate the spatial arrangement of atoms.

7.2.4 Numerical Example

Consider the reaction between hydrogen (H_2) and oxygen (O_2) to form water (H_2O). The unbalanced equation is:

$$H_2 + O_2 \longrightarrow H_2O$$

Let's find the amount of water produced when 3 moles of hydrogen react. Using stoichiometry:

$$\text{Moles of water} = 3\,\text{mol} \times \frac{1}{2} = 1.5\,\text{mol}$$

7.3   Stoichiometry and Mole Calculations

Stoichiometry is a branch of chemistry that deals with the quantitative relationships between reactants and products in chemical reactions. Mole calculations are an essential aspect of stoichiometry, allowing chemists to relate the amounts of substances involved. Let's explore stoichiometry and mole calculations through examples.

7.3.1   Example 1: Reaction Between Hydrogen and Oxygen

Consider the reaction between hydrogen (H_2) and oxygen (O_2) to form water (H_2O). The balanced equation is:

$$2\,H_2 + O_2 \longrightarrow 2\,H_2O$$

This equation indicates that two moles of hydrogen react with one mole of oxygen to produce two moles of water.

Numerical Example

Suppose we have 3 moles of hydrogen. To find the amount of oxygen needed and the moles of water produced:

$$\text{Moles of oxygen} = 3\,\text{mol} \times \frac{1}{2} = 1.5\,\text{mol}$$

$$\text{Moles of water} = 3\,\text{mol} \times 2 = 6\,\text{mol}$$

7.3.2   Example 2: Formation of Ammonia

Consider the synthesis of ammonia (NH_3) from nitrogen (N_2) and hydrogen (H_2). The balanced equation is:

$$N_2 + 3\,H_2 \longrightarrow 2\,NH_3$$

This equation indicates that one mole of nitrogen reacts with three moles of hydrogen to produce two moles of ammonia.

Numerical Example

Suppose we have 2 moles of nitrogen. To find the moles of hydrogen needed and the moles of ammonia produced:

$$\text{Moles of hydrogen} = 2\,\text{mol} \times 3 = 6\,\text{mol}$$

$$\text{Moles of ammonia} = 2\,\text{mol} \times 2 = 4\,\text{mol}$$

7.3.3 Bonding Diagrams

Let's use to draw the bonding diagram for ammonia (NH_3):

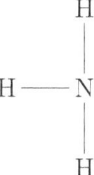

In this diagram, the lines represent bonds, and the angles indicate the spatial arrangement of atoms.

Chapter 8

Periodic Table Applications

8.1 Predicting Element Properties

The periodic table is a powerful tool for predicting the properties of elements based on their positions. Let's explore how certain trends and patterns in the periodic table can help us anticipate various element properties.

8.1.1 Electronegativity Trends

Electronegativity, the tendency of an atom to attract electrons in a chemical bond, exhibits trends across the periodic table. Generally, electronegativity increases from left to right across a period and decreases from top to bottom within a group.

Numerical Example

Consider the electronegativity trend for halogens (Group 17 elements). Fluorine (F) is more electronegative than chlorine (Cl). This trend helps predict that other halogens will have decreasing electronegativities down the group.

8.1.2   Atomic Radius Trends

The atomic radius, or the size of an atom, also follows trends on the periodic table. Atomic radius tends to decrease from left to right across a period and increase from top to bottom within a group.

Numerical Example

Compare the atomic radii of alkali metals (Group 1 elements). Potassium (K) has a larger atomic radius than lithium (Li). This trend allows us to predict that other alkali metals will have increasing atomic radii down the group.

8.1.3   Bonding Diagrams

Let's use to draw the bonding diagram for a diatomic molecule of oxygen (O_2):

$$O = O$$

In this diagram, the double bond between oxygen atoms is represented.

8.2   Chemical Analysis and Spectroscopy

Chemical analysis and spectroscopy are powerful techniques used to identify and quantify elements and compounds. In this section, we'll explore some of the methods and applications of these analytical tools.

8.2.1   Mass Spectrometry

Mass spectrometry is a technique that measures the mass-to-charge ratio of ions. It is widely used in chemical analysis and helps identify the composition of a sample.

Working Example

Consider the mass spectrum of methane (CH_4):

$$
\begin{array}{c}
\text{H} \\
| \\
\text{H} \!-\!\! \text{C} \!-\!\! \text{H} \\
| \\
\text{H}
\end{array}
$$

The mass spectrum will show peaks corresponding to different isotopes of carbon and hydrogen, allowing for precise identification.

8.2.2 Nuclear Magnetic Resonance (NMR)

NMR spectroscopy is based on the magnetic properties of atomic nuclei. It provides valuable information about molecular structure and dynamics.

Numerical Example

The NMR spectrum of ethanol (C_2H_5OH) reveals distinctive peaks for different hydrogen environments, aiding in the determination of molecular structure.

8.2.3 UV-Visible Spectroscopy

UV-Visible spectroscopy measures the absorption of light in the ultraviolet and visible regions. It is commonly used to study electronic transitions.

Bonding Diagram

Let's use to draw the bonding diagram for a molecule with conjugated double bonds:

This represents a molecule with extended conjugation.

8.3   Periodic Trends in Everyday Life

The periodic table offers insights into various elements and their properties, impacting our daily lives in surprising ways.

8.3.1   Electronegativity and Bonding

Working Example

Consider the electronegativity difference between chlorine (Cl) and hydrogen (H). The electronegativity trend helps predict the nature of the bond in HCl.

$$H \textrm{---} Cl$$

Chlorine, being more electronegative, attracts the shared electrons, resulting in a polar covalent bond.

8.3.2   Ionization Energy and Reactivity

Numerical Example

The ionization energy of alkali metals, like sodium (Na), is relatively low. This makes them highly reactive. For instance:

$$Na \longrightarrow Na^+ + e^-$$

Sodium readily loses an electron to achieve a stable electron configuration.

8.3.3   Atomic Radius and Properties of Metals

Bonding Diagram

Using , let's visualize the metallic bonding in sodium (Na):

$$N\underset{e}{a}^+$$

Metallic bonding involves a sea of delocalized electrons surrounding positively charged metal ions.

Chapter 9

Advanced Topics in Inorganic Chemistry

9.1 Transition Metal Coordination Chemistry

Transition metal coordination compounds play a crucial role in inorganic chemistry. These compounds exhibit unique structures and bonding behaviors.

9.1.1 Introduction to Coordination Compounds

Coordination compounds consist of a central metal ion or atom surrounded by ligands. Ligands are molecules or ions that can donate electron pairs to form coordinate bonds.

Coordination Number and Geometry

The coordination number represents the number of ligands bonded to the central metal. Different coordination numbers lead to various geometries. For instance, octahedral coordination (CN = 6) results in a symmetrical arrangement.

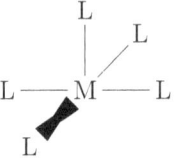

9.1.2 Working Example: Complex Formation

Consider the formation of a coordination complex between copper (Cu^{2+}) and ammonia (NH_3). The balanced chemical equation is:

$$Cu^{2+} + 4\,NH_3 \longrightarrow [Cu(NH_3)_4]^{2+}$$

The coordination number of copper in the complex is 4, leading to a tetrahedral geometry.

9.1.3 Electronic Structure and Bonding

Crystal Field Theory

Crystal Field Theory helps explain the color of transition metal complexes. When ligands approach, the d-orbitals split into higher and lower energy levels.

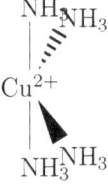

Numerical Example

Consider a coordination complex with an unpaired electron in its d-orbital. This complex exhibits magnetic behavior. For example:

$$[Fe(H_2O)_6]^{3+}$$

The unpaired electron leads to magnetic moments.

9.2 Organometallic Compounds

Organometallic compounds, featuring direct metal-carbon bonds, play a pivotal role in modern inorganic chemistry. These compounds exhibit unique reactivity and catalytic properties.

9.2.1 Introduction to Organometallic Compounds

Organometallic compounds involve metal atoms directly bonded to carbon atoms. The metal-carbon bond can be either sigma (σ) or pi (π).

Classification

Organometallic compounds are classified based on the nature of the metal-carbon bond and the coordination environment of the metal center.

Representation

The representation of organometallic compounds often involves the use of structural formulas:

$$M \text{---} C$$

9.2.2 Working Example: Ferrocene

Ferrocene ($Fe(C_5H_5)_2$) is a classic example of an organometallic compound. It consists of a sandwich-like structure with an iron (Fe^{2+}) center.

$$Cp_2Fe$$

9.2.3 Bonding and Reactivity

Bonding in π-Complexes

Organometallic compounds often form π-complexes due to the presence of multiple bonds between the metal and carbon atoms.

$$M\!\!-\!\!\!-C$$

9.2.4   Catalytic Applications

Homogeneous Catalysis

Organometallic compounds find applications in homogeneous catalysis, facilitating various chemical transformations.

$$2\,PhMgBr + TiCl_4 \xrightarrow{\text{THF}} Ti(Ph)_2Cl_2 + 2\,MgBrCl$$

9.3   Bioinorganic Chemistry

Bioinorganic chemistry explores the role of metals in biological systems, uncovering the intricate interplay between metal ions and living organisms.

9.3.1   Introduction to Bioinorganic Chemistry

Bioinorganic chemistry delves into the vital functions of metal ions in biological processes, spanning metalloproteins, metalloenzymes, and metal-containing cofactors.

Metalloproteins

Metalloproteins incorporate metal ions within their structures, contributing to diverse biological functions.

Metalloenzymes

Metalloenzymes harness the catalytic power of metal ions, enabling essential biochemical reactions.

9.3.2 Working Example: Hemoglobin

Hemoglobin, a quintessential metalloprotein, illustrates the critical role of metals in biological systems. Its heme group contains an iron ion that facilitates oxygen transport.

$$Fe^{3+}\text{porphyrin ring}$$

9.3.3 Metal Homeostasis

Living organisms tightly regulate metal concentrations to maintain cellular functions and prevent metal toxicity.

Metal Transporters

Specialized metal transporters facilitate the controlled movement of metal ions across cellular membranes.

$$Cu^{2+} \text{ (outside)} \xrightarrow{\text{transporter}} Cu^{2+} \text{ (inside)}$$

9.3.4 Biomineralization

Biomineralization involves the controlled formation of mineral structures within living organisms, often mediated by proteins.

$$Ca^{2+} - \text{protein matrix} \xrightarrow{\text{biomineralization}} \text{calcified structure}$$

9.3.5 Medical Applications

Understanding bioinorganic chemistry is crucial for developing medical interventions, including metal-based drugs and diagnostic agents.

Cisplatin

Cisplatin, a platinum-based drug, exemplifies the use of metal complexes in cancer treatment.

$$[Pt(NH_3)_2Cl_2] \xrightarrow{\text{cell uptake}} Pt{-}DNA\text{ adduct} \xrightarrow{\text{cell apoptosis}} \text{killed cancer cell}$$

Chapter 10

Quantum Mechanics and the Periodic Table

10.1 Wave-Particle Duality

Wave-particle duality is a fundamental concept in quantum mechanics that describes the dual nature of particles, such as electrons and photons. According to this principle, particles exhibit both wave-like and particle-like characteristics, depending on the experimental conditions.

10.1.1 De Broglie Wavelength

The de Broglie wavelength (λ) is a concept that associates a wavelength with any particle. It is given by the equation:

$$\lambda = \frac{h}{p}$$

where h is the Planck constant and p is the momentum of the particle. This wavelength is significant in understanding the wave-like behavior of particles.

10.1.2 Wave-Particle Duality Experiments

Several experiments demonstrate the wave-particle duality of particles. One notable experiment is the double-slit experiment. In this experiment, particles such as electrons are sent through two slits, creating an interference pattern on the screen behind them. This pattern is characteristic of waves, confirming the wave-like nature of particles.

10.1.3 Application to Quantum Mechanics

Wave-particle duality has profound implications for the behavior of particles at the quantum level. It is a cornerstone in the development of quantum mechanics, providing insights into phenomena like the Heisenberg Uncertainty Principle and the quantization of energy levels.

10.1.4 Bonding and Wave-Particle Duality

In the context of chemistry, understanding wave-particle duality is crucial for comprehending chemical bonding. The electronic structure of atoms, including the concept of atomic orbitals and hybridization, can be explained more thoroughly through the lens of wave-particle duality.

10.1.5 Structures

Using , we can represent molecular structures in a clear and concise manner. For example, the molecular structure of water (H_2O) is represented as:

$$H \text{——} O \text{——} H$$

This representation allows for a visual understanding of the arrangement of atoms within the molecule.

10.2 Quantum Numbers

In the realm of quantum mechanics, quantum numbers are essential parameters that define the properties of electrons within an atom. They provide a framework for understanding the distribution of electrons in atomic orbitals and contribute to the overall structure of the periodic table.

10.2.1 Principal Quantum Number (n)

The principal quantum number, denoted as n, determines the energy level of an electron and the size of its orbital. For a given element, n can take positive integer values (1, 2, 3, ...) corresponding to different electron shells.

10.2.2 Angular Momentum Quantum Number (l)

The angular momentum quantum number, l, specifies the shape of the orbital. It ranges from 0 to $n - 1$ and is often represented by letters: s (0), p (1), d (2), f (3), and so on. Each letter corresponds to a specific orbital shape.

10.2.3 Magnetic Quantum Number (m_l)

The magnetic quantum number, m_l, determines the orientation of the orbital in space. Its values range from $-l$ to l, including zero. For example, for an s orbital ($l = 0$), m_l can only be 0.

10.2.4 Spin Quantum Number (m_s)

The spin quantum number, m_s, describes the intrinsic spin of an electron. It can have two values: $+\frac{1}{2}$ (spin-up) or $-\frac{1}{2}$ (spin-down). This quantum number is crucial for the Pauli Exclusion Principle.

10.2.5 Representation with Matrices

Quantum numbers can be represented using matrices. For example, the matrix representation of the angular momentum operator (L_z) in the z-direction is given by:

$$L_z = -i\hbar \frac{d}{d\phi}$$

where $\hbar$ is the reduced Planck constant and ϕ is the azimuthal angle.

10.2.6 Structures

Using , we can visually represent electron configurations and orbital structures. For instance, the electron configuration of carbon ($Z = 6$) can be represented as:

$$1s^2 2s^2 2p^2$$

This notation provides a concise way to express the distribution of electrons in different orbitals.

10.3 Quantum Mechanical Model of the Atom

The quantum mechanical model of the atom is a crucial framework that revolutionized our understanding of atomic structure. Developed in the early 20th century, it replaced the classical model with a more sophisticated and accurate description based on the principles of quantum mechanics.

10.3.1 Key Concepts

Wave-Particle Duality

One of the fundamental concepts is the wave-particle duality, which suggests that particles, such as electrons, exhibit both wave and particle properties. This duality is expressed by the de Broglie wavelength equation:

$$\lambda = \frac{h}{p}$$

where λ is the wavelength, h is the Planck constant, and p is the momentum.

Heisenberg Uncertainty Principle

The Heisenberg Uncertainty Principle states that it is impossible to simultaneously know both the exact position and momentum of a particle. Mathematically, it can be expressed as:

$$\Delta x \cdot \Delta p \geq \frac{\hbar}{2}$$

where Δx is the uncertainty in position, Δp is the uncertainty in momentum, and $\hbar$ is the reduced Planck constant.

Quantum Numbers

Quantum numbers, including the principal quantum number (n), angular momentum quantum number (l), magnetic quantum number (m_l), and spin quantum number (m_s), are integral to the quantum mechanical model. They define the allowed energy states and orbital characteristics of electrons.

10.3.2 Matrix Representation

The quantum mechanical model often utilizes matrix representations for operators corresponding to physical observables. For example, the matrix representation of the Hamiltonian operator ($\hat{H}$) is given by:

$$\hat{H} = -\frac{\hbar^2}{2m}\nabla^2 + V$$

where m is the mass of the particle and V is the potential energy.

10.3.3 Chemfig Structures

The electronic structure of atoms and molecules can be visually represented using . For instance, the Lewis structure of water (H_2O) can be drawn as:

This notation provides a simple yet powerful representation of molecular geometry.

10.3.4   Applications

The quantum mechanical model is indispensable for predicting the behavior of electrons in atoms, explaining atomic spectra, and understanding chemical bonding. It forms the basis for modern quantum chemistry and contributes significantly to the arrangement of elements in the periodic table.

Chapter 11

Nuclear Chemistry and the Periodic Table

11.1 Radioactive Decay

The phenomenon of radioactive decay plays a crucial role in nuclear chemistry and has significant implications for understanding the behavior of certain isotopes. This section explores the principles of radioactive decay, its mathematical description, and its relevance to the periodic table.

11.1.1 Introduction to Radioactive Decay

Radioactive decay is the spontaneous process through which unstable atomic nuclei lose energy by emitting radiation. This process is common among certain isotopes, particularly those with an excess of either protons or neutrons in their nuclei. The decay is governed by the inherent instability of these nuclei.

11.1.2 Mathematical Description

The decay of a radioactive substance can be mathematically described using the decay equation:

$$N(t) = N_0 \cdot e^{-\lambda t}$$

where:

$N(t)$: the quantity of radioactive substance at time t

N_0 : the initial quantity of the substance

λ : the decay constant

This equation helps determine the remaining quantity of a substance over time.

11.1.3   Types of Radioactive Decay

There are several types of radioactive decay, including alpha decay, beta decay, and gamma decay. Each type involves the emission of specific particles or electromagnetic radiation.

Alpha Decay

Alpha decay occurs when an unstable nucleus emits an alpha particle (α), consisting of two protons and two neutrons. This process reduces the atomic number by 2 and the mass number by 4.

$$^{238}_{92}U \longrightarrow ^{234}_{90}Th + ^{4}_{2}\alpha$$

Beta Decay

Beta decay involves the transformation of a neutron into a proton or vice versa, accompanied by the emission of a beta particle (β).

$$^{14}_{6}C \longrightarrow ^{14}_{7}N + ^{0}_{-1}\beta$$

Gamma Decay

Gamma decay releases gamma rays (γ), high-energy electromagnetic radiation, without changing the atomic or mass numbers.

$$^{60}_{27}\text{Co}^* \longrightarrow {}^{60}_{27}\text{Co} + {}^{0}_{0}\gamma$$

11.1.4 Applications

Understanding radioactive decay is crucial in various fields, including archaeology (carbon dating), medicine (radioactive tracers), and energy production (nuclear power).

11.2 Nuclear Reactions

Nuclear reactions involve changes in the structure of atomic nuclei. These reactions play a crucial role in nuclear chemistry and contribute to various scientific and technological applications. In this section, we will explore the basics of nuclear reactions, their types, and their significance.

11.2.1 Types of Nuclear Reactions

Nuclear reactions can be categorized into several types, including:

- **Alpha Decay:** The emission of an alpha particle (α) from a nucleus.

- **Beta Decay:** The conversion of a neutron into a proton or vice versa, with the emission of a beta particle (β).

- **Gamma Decay:** The emission of a gamma ray (γ), often accompanying alpha or beta decay, to achieve a more stable nuclear state.

- **Nuclear Fusion:** The combining of two lighter atomic nuclei to form a heavier nucleus.

- **Nuclear Fission:** The splitting of a heavy atomic nucleus into two or more lighter nuclei, accompanied by the release of energy.

11.2.2 Representation of Nuclear Reactions

Nuclear reactions can be represented using nuclear equations. For instance, the alpha decay of uranium-238 ($^{238}_{92}U$) can be expressed as:

$$^{238}_{92}U \xrightarrow{\alpha} {}^{234}_{90}Th + {}^{4}_{2}\alpha$$

This representation highlights the transformation of uranium-238 into thorium-234 through the emission of an alpha particle.

11.2.3 Applications of Nuclear Reactions

Nuclear reactions find applications in various fields, such as:

- **Nuclear Power Generation:** Utilizing controlled nuclear fission reactions to generate electricity.

- **Medical Imaging:** Employing radioactive isotopes for diagnostic imaging techniques like positron emission tomography (PET).

- **Radiation Therapy:** Using targeted nuclear reactions for cancer treatment.

- **Isotope Production:** Creating specific isotopes for medical, industrial, or research purposes.

Understanding nuclear reactions is crucial for advancing nuclear science and technology, with implications for energy production, medicine, and scientific research.

11.3 Isotopes and Applications

Isotopes, variants of a chemical element with the same number of protons but different numbers of neutrons, play a crucial role in nuclear chemistry and various applications. This section explores the significance of isotopes and their diverse applications.

11.3.1 Understanding Isotopes

Isotopes are atoms of the same element with different masses due to variations in the number of neutrons. For example, carbon has two stable isotopes: carbon-12 (^{12}C) and carbon-13 (^{13}C). The differences in isotopic composition can affect the chemical and physical properties of substances.

11.3.2 Applications of Isotopes

Medical Imaging

One of the notable applications of isotopes is in medical imaging. Radioactive isotopes, such as technetium-99m (^{99m}Tc), are widely used in diagnostic procedures like single-photon emission computed tomography (SPECT) scans. The decay of these isotopes emits gamma rays, aiding in visualizing internal structures within the body.

Carbon Dating

Isotopes, particularly carbon-14 (^{14}C), find application in archaeology and geology for carbon dating. By measuring the decay of carbon-14 in organic materials, scientists can determine the age of ancient artifacts, fossils, and geological samples.

Radioactive Tracers

Isotopes with suitable decay properties serve as radioactive tracers in scientific research. For instance, iodine-131 (^{131}I) is used in medical studies to trace the path of substances within the body.

11.3.3   Isotopic Notation

The notation for isotopes includes the element symbol, mass number (sum of protons and neutrons), and atomic number. For example, the isotopic notation for uranium-235 ($^{235}_{92}U$) signifies an element with 92 protons, 235 total nucleons (protons and neutrons combined).

11.3.4   Isotopic Composition

The isotopic composition of an element is expressed as a percentage of each isotope's abundance in a given sample. This information is vital for understanding the behavior of isotopes in chemical reactions and natural processes.

Appendix A

Common Chemical

Constants and Units

This appendix provides a comprehensive list of common chemical constants and units essential for understanding and performing chemical calculations. Familiarity with these constants is crucial in various scientific disciplines, including chemistry, physics, and engineering.

A.1   Fundamental Constants

- Avogadro's Number (N_A): The number of atoms or molecules in one mole, approximately $6.022 \times 10^{23}\,\mathrm{mol}^{-1}$.

- Boltzmann's Constant (k): The constant relating the average kinetic energy of particles in a gas to the temperature, approximately $8.617 \times 10^{-5}\,\mathrm{eV\ K}^{-1}$.

- Speed of Light (c): The speed at which light propagates in a vacuum, approximately $3.00 \times 10^8\,\mathrm{m/s}$.

A.2   Common Units

A.2.1   SI Units

- Meter (m): The base unit of length in the International System of Units.

- Kilogram (kg): The base unit of mass in the International System of Units.

- Second (s): The base unit of time in the International System of Units.

- Ampere (A): The base unit of electric current in the International System of Units.

A.2.2   Derived Units

- Joule (J): The unit of energy, equivalent to one kilogram meter squared per second squared ($kg\ m^2/s^2$).

- Coulomb (C): The unit of electric charge.

- Kelvin (K): The unit of temperature.

A.3   Mathematical Constants

- Euler's Number (e): The mathematical constant representing the base of natural logarithms, approximately 2.71828.

- Pi (π): The ratio of a circle's circumference to its diameter, approximately 3.14159.

Appendix B

Glossary of Terms

This glossary provides concise definitions of key terms and concepts encountered throughout the book. It serves as a valuable reference for readers seeking clarification on specific terminology related to chemistry, the periodic table, and associated topics.

B.1   Selected Terms

- **Atom:** The basic unit of a chemical element, consisting of a nucleus of protons and neutrons, with electrons in orbit around the nucleus.

- **Bonding:** The process of joining two or more atoms to form molecules or compounds, often involving the sharing or transfer of electrons.

- **Isotope:** Atoms of the same element with the same number of protons but different numbers of neutrons, resulting in variations in atomic mass.

- **Mole:** A unit of measurement in chemistry representing an amount of substance containing the same number of entities as there are atoms in 12 grams of carbon-12.

- **Periodic Table:** A tabular arrangement of chemical elements based on their atomic number, electron configuration, and recurring chemical properties.

- **Quantum Mechanics:** A branch of physics that describes the behavior of matter and energy at the atomic and subatomic levels, incorporating principles of wave-particle duality.

- **Stoichiometry:** The calculation of reactants and products in chemical reactions, often involving the use of balanced chemical equations.

B.2   Advanced Concepts

- **Coordination Chemistry:** The study of coordination compounds, which consist of a central metal atom or ion bonded to surrounding ligands.

- **Nuclear Chemistry:** The branch of chemistry that deals with the properties and behavior of radioactive elements and their interactions with non-radioactive elements.

- **Quantum Numbers:** Parameters that describe the state of an electron in an atom, including principal quantum number, azimuthal quantum number, magnetic quantum number, and spin quantum number.

- **Wave-Particle Duality:** The concept in quantum mechanics that particles, such as electrons, exhibit both wave-like and particle-like properties depending on the experimental conditions.

www.ingramcontent.com/pod-product-compliance
Lightning Source LLC
Chambersburg PA
CBHW082245310526
45795CB00015B/2998